SCIENCE WORKSHOP

CLOCKS · SCALES & MEASUREMENTS

Pam Robson

GLOUCESTER PRESS
NEW YORK CHICAGO LONDON TORONTO SYDNEY

First published in
the United States in 1993 by
Franklin Watts
95 Madison Avenue
New York, NY 10016

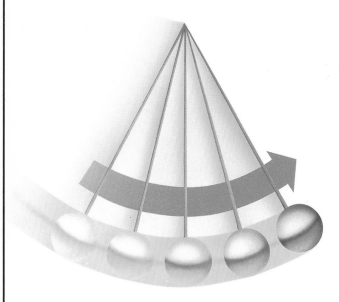

Library of Congress Cataloging-in-Publication Data

Robson, Pam.
    Clocks, scales & measurements / Pam Robson.
    p.   cm. — (Science workshop)
    Includes index.
    Summary: Presents information and activities
related to the measurement of time, weight, area,
density, and other aspects of the world around us.
    ISBN 0-531-17419-0
    1. Physical instruments—Juvenile literature. 2.
Physical measurements— Experiments— Juvenile
literature. [1. Measurement—Experiments. 2.
Experiments.]
    I. Title. II. Title : Clocks, scales, and
measurements. III. Series
    QC55.R6   1993
    530.8'078—dc20
    93-19873 CIP AC

CONTENTS

Design	David West
	Children's Book Design
Designer	Steve Woosnam-Savage
Editor	Suzanne Melia
Picture Researcher	Emma Krikler
Illustrators	Ian Thompson
Consultant	Bryson Gore

PHOTOCREDITS

All the photographs in this book are by Roger Vlitos apart
from pages; 6 top: Science Photo Library; 8 top, 22 top & 24
top: Eye Ubiquitous; 12 top right: Reproduced by permission
of the Trustees of the Science Museum; 18 top & 20 top:
Spectrum Colour Library.

INTRODUCTION

Listen for the questions "How many?" and "How much?" During an average day, you are likely to hear them quite a few times. The answers to both questions use numbers. But the first question is answered by counting and the second by measuring. We use measurements every day of our lives. Each time we look at a clock, we are reading a measurement of time. Whenever we travel, short or long distances, we need to know how long the journey will take, how far away our destination is and in which direction we are traveling. Buying food and clothes also involves measurement and a familiarity with the scales and units being used. Standard measuring tools and scales must be used around the world for the system to work. The British system (feet and inches) was developed around the 1200s, but most countries now use the metric system (meters and centimeters). A standard meter is one-ten millionth of a straight line distance from the North Pole to the Equator. If you think this sounds fascinating, read on to find out how scales and units of measurement were developed and how they are used. You'll realize what a chaotic world it would be without them!

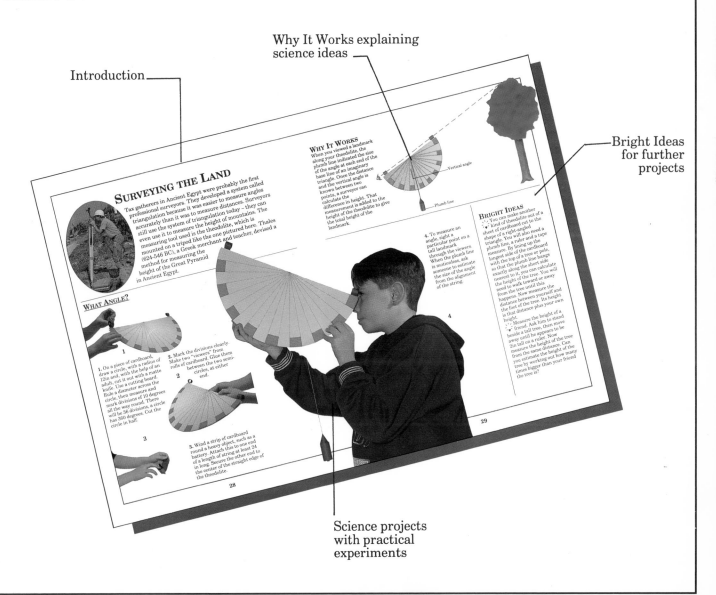

Introduction

Why It Works explaining science ideas

Bright Ideas for further projects

Science projects with practical experiments

THE WORKSHOP

A science workshop is a place to test ideas, perform experiments and make discoveries. To prove many scientific facts, you don't need a lot of fancy equipment. In fact, everything you need for a basic workshop can be found around your home or school. Read through these pages, and then use your imagination to add to your "home laboratory." As you work your way through this book, you should enjoy completing the projects and seeing your models work. Remember, though, that from a scientific point of view, these projects are just the starting point. For example, when you finish making the egg-timer on pages 10/11, ask your own questions like "What would happen if I used sand?", "Would a larger hole let the sand flow faster?", and so on. Also, by sharing ideas, you will learn more. Experimenting with equipment, as well as with ideas, will give you the most accurate results. In this way you will build up your workshop as you go along.

MAKING MODELS

Before you begin, read through all the steps. Then, make a list of the things you need and collect them together. Next, think about the project so that you have a clear idea of what you are about to do. Finally, take your time in putting the pieces together. You will find that your projects work best if you wait while glue or paint dries. If something goes wrong, retrace your steps. And, if you can't fix it, start over again. Every scientist makes mistakes, but the best ones know when to begin again!

SAFETY WARNINGS

Make sure that an adult knows what you are doing at all times. Cutting the top off a plastic bottle can be difficult and dangerous if you use sharp scissors. Ask an adult to do this for you. Always be careful with balloons and plastic bags. Never cover your face with them. If you spill any water, wipe it up right away. Slippery surfaces are dangerous. Clean up after you have finished.

GENERAL TIPS

There are at least two parts to every experiment: experimenting with materials and testing a science "fact." If you don't have all the materials, experiment with others instead. For example, if you can't find any glass bottles, use some wine glasses instead, if your parent approves. Once you've finished experimenting, read your notes thoroughly and think about what happened, evaluating your measurements and observations. What conclusions can you draw from your results?

EXPERIMENTING

Always conduct a "controlled experiment." This means changing one thing at a time for each stage of an experiment. In this way, you can always tell which change caused a different result. As you go along, record what you see. Ask questions such as "why?", "how?" and "what if?". Then test your model and write down the answers you arrive at. Compare your results to those of your classmates or friends.

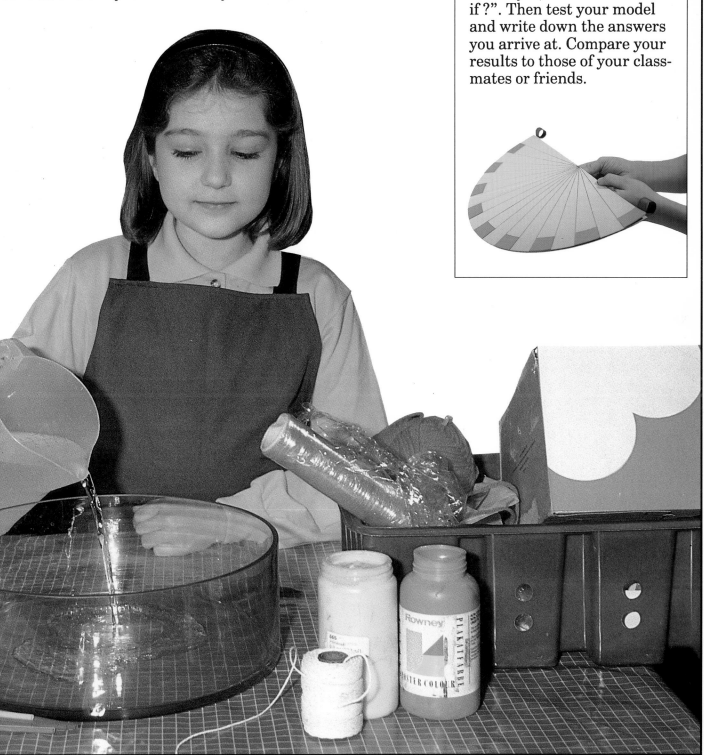

CALENDARS

Before the invention of the clock, people had to rely on nature's timekeepers - the Sun, the Moon and the stars. The daily movement of the Sun across the sky provided the simplest unit, the solar day. The time period of a year was estimated by watching the seasons, and the constancy of the lunar cycle led to the division of each year into months. Traditionally, calendars were controlled by priests. They were devised either by counting days or by following the phases of the Moon. Nowadays, the Gregorian calendar is most common. This was worked out by Pope Gregory XIII in the 1580s.

DAY BY DAY

1. Cut out two circles of cardboard. The largest should be 12 in across, the other 11in across. Stick one on top of another and divide into 12 equal pieces to indicate the months.

2. Cut out 12 paper circles of 0.5in across. Find out the number of days for each month and write them around the edge of each small circle. Now stick them in order around the large circle. The first day of the month should be nearest the edge as shown.

1

2

3. Cut another cardboard circle 10.5in across. Cut a hole, radius 0.5in, to correspond with the position of the paper circles. Cover the hole with stiff, transparent plastic. Attach a red arrow marker, as shown.

3

4. Cut out a cardboard circle, radius 0.5in. Carefully make a tiny "window," to view the date through. Position it over the 0.5in hole, and fix it to the plastic with a paper fastener, so it turns.

4

5

5. Decorate your calendar before joining the separate sections together. Position the smaller circle centrally over the larger circle and join them together with a paper fastener. Rotate your calendar until it is set on the correct day for the current month.

WHY IT WORKS

A calendar is a system of time measurement.
Our calendar is based on the movements
of the planets. The Earth rotates
once every 24 hours, or once a
day. The Moon orbits the
Earth once every month,
and the Earth takes 365
days, or 1 year to
orbit the Sun.

BRIGHT IDEAS

Design a tally system for
marking off the days,
using symbols instead of
numbers. Can you think of a
way to group a certain
number of tally marks
together to make a single tally
mark that represents that
larger number? We say 7 days
equal one whole week. Days
and weeks are units of time.

Find out how and when
different cultures
celebrate New Year. How do
their calendars work?

6. View the date
through the
"window." Remember
to turn the small wheel
daily. Each revolution of the
0.5in circle is equivalent to
one month, as represented on
the calendar. For each new
month rotate the large circle.

7

PLOTTING THE STARS

Astronomy is the study of the stars, planets and other objects in the universe. For centuries, astronomers have striven to learn more about our Universe. Through observation and careful measurement, using scientific tools like the telescope, we now know that the Sun is the center of our solar system. Accurate measurement of star distance is a science developed over the centuries by astronomers like Tycho Brahe (1546-1601). Centuries ago, sailors calculated time at night by observing the movements of star clusters near the fixed Pole Star. Watches aboard ship were timed from the position of these constellations in the sky.

STAR TIME

1

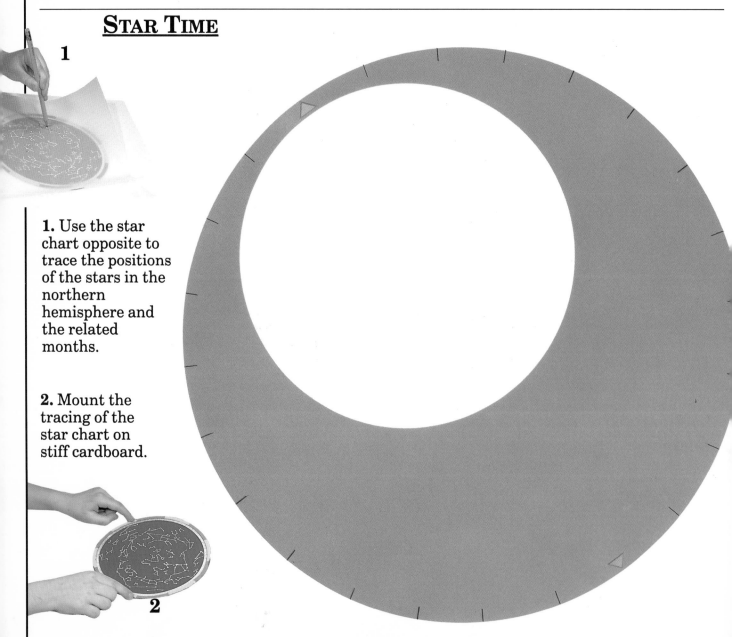

1. Use the star chart opposite to trace the positions of the stars in the northern hemisphere and the related months.

2. Mount the tracing of the star chart on stiff cardboard.

2

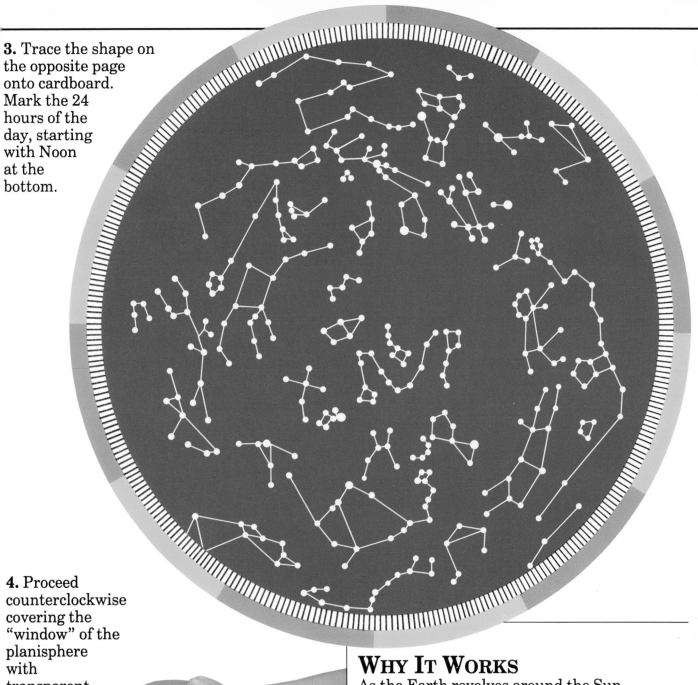

3. Trace the shape on the opposite page onto cardboard. Mark the 24 hours of the day, starting with Noon at the bottom.

4. Proceed counterclockwise covering the "window" of the planisphere with transparent plastic. Pin it to the starchart through the center.

4

3

5. On a starry night, rotate the planisphere until the time of day, marked on the edge, is lined up with the appropriate month at the bottom of the star chart. Compare what you see with the stars in the sky.

WHY IT WORKS

As the Earth revolves around the Sun, different constellations of stars appear and disappear in a continous cycle that can be observed. This planisphere shows where these groups of stars can be seen at any given time of the year, in the northern hemisphere. By matching up the time of day with the date, you can view the stars that should be visible in the night sky through the cut out "window." The planisphere should be held up and viewed from underneath. The stars visible through the window should match those in the sky. The Sun is a star. It is the only star close enough to look like a ball. The other billions of stars are so far away, they appear to be pinpoints of light.

MEASURING ROTATION

The Earth revolves, on its own tilted axis, once every 24 hours. The Sun appears to rise and set. The rotation of the Earth was demonstrated in 1851, by J.B.L. Foucault, a French physicist. He realized that although a swinging pendulum seems to change direction throughout the day, it is really the Earth that rotates beneath it while the swing of the pendulum tries to remain constant. By studying the movement of the planets, astronomers have noted a gradual slowing of the Earth's rotation. They have calculated that the length of a day increases by about 1.7 milliseconds per year.

IN A SPIN

1

1. Fill a circular bowl with clean, cold water. Cut out a circle of cardboard, slightly larger than the diameter of the bowl. Draw a straight line across, marking each end with an arrow.

4

2. Find a flat surface, free of vibration. Position the cardboard on the surface, then stand the bowl of water on top. Leave it to stand until the surface of the water is still.

4. Rest the card circle on top of the bowl. Line up the slit with the marks on the card underneath the bowl, as shown. Prepare some powdered cork and keep in a dry container.

3

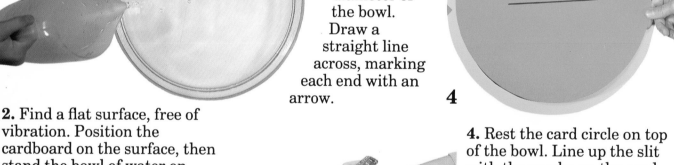

3. Cut another circle of cardboard, slightly larger than the bowl. Rule a line across the center. Using a matte knife, on a board, cut out a slit approximately 0.1in wide, as shown.

2

5

5. If the water is still, gently sprinkle the cork powder over the slit in the cardboard. It will fall through to make a narrow line of powder on the surface of the water.

WHY IT WORKS

The line of cork dust appears to have rotated after the bowl has been standing for a number of hours. In reality, the Earth and the bowl have rotated beneath the line of dust, which remained stationary. This is because the Earth travels through 360 degrees each time a rotation is completed. Every place on Earth travels through 15 degrees each hour of the day. Artificial time zones have been created around the world to accommodate time differences.

Earth

Direction of rotation

BRIGHT IDEAS

Make a pendulum by pushing a needle right through an apple and suspending it inside an empty box with thread. The thread should be just the right length to let the needle sticking out of the apple touch a plate underneath. Sprinkle some fine salt on the plate and start the pendulum swinging. Turn the plate slowly and watch the grooves form as the pendulum swings in the same plane and the plate rotates. The plate represents the Earth as it rotated in Foucault's pendulum experiment.

6. Do not touch the bowl. Leave it undisturbed for several hours. You will find, after some time, that the line of dust appears to have rotated and is no longer in line with the arrows.

6

ESTIMATING TIME

Hours are artificial units of time, first introduced by the Ancient Egyptians when they observed that shadows follow a similar pattern of movement each day. Shadow clocks and sundials were early time-measurement tools. In the 14th century, the sandglass, or hourglass, was popular, but it could only be used to estimate periods of time, varying from minutes to hours. It could not indicate the time of day. On board ship, a four-hour glass timed "watch" for the crew, until John Harrison invented the more accurate chronometer in 1735. This also calculated longitude and latitude. Early sandglasses were filled with powdered eggshell or marble dust.

TIME IS RUNNING OUT

1

1. Wash and dry two small bottles thoroughly. Make an open-ended cylinder of cardboard and slide it over the top of one bottle. Cut out a disc of cardboard to fit inside and make a hole in its center.

2

2. Make sure the other bottle is absolutely dry. Now, carefully pour a measured amount of salt into it.

3

3. Position the empty bottle on top of the salt-filled bottle by sliding the cardboard cylinder over the neck of the lower bottle.

4

4. Check that the cardboard "seal" round the middle is secure. Carefully turn your timer over and observe what happens.

5. You can estimate the time taken for the salt to slide into the lower half of the timer by marking the side of the bottle with evenly spaced divisions. Use a stop watch to check exactly how long it takes for all the salt to slide to the bottom.

5

WHY IT WORKS

The upper vessel of the timer holds just enough salt to run through a hole, of a given size, in a given period of time. The force of gravity pulls the salt down through the hole and into the bottom container. The salt grains must be absolutely dry, so they don't stick. The size of the hole between the two vessels will determine the speed at which the grains will flow, but once established, the rate of flow will not vary. The total period of time depends on the quantity of salt. Salt—

BRIGHT IDEAS

Remove the regulator from between the two bottles and replace it with another, in which a hole of a different diameter has been made. Repeat this exercise a number of times, changing the size of the hole each time. What do you discover?

Design a sandglass that will run for exactly 3 minutes - use it as an eggtimer when you boil an egg.

Can you design another kind of sandglass that runs for a much longer period of time? Shape a funnel from cardboard, and insert it into the neck of a measuring container. Fill the funnel with sand and time how long it takes to pour through into the container below. Standardize your method of reading the scale - the top of the mound of sand will be concave.

Funnel

Sand—

Scale—

MEASURING TIME

The accurate measurement of small units of time was not possible until 1583. The Italian physicist Galileo (1564-1642) observed the swinging of a lamp in Pisa cathedral, and by timing its motion against his own pulse, he concluded that a free pendulum was isochronous – every swing, wide or narrow, took the same time. The Dutch scientist Christiaan Huygens (1629-1695) constructed the first mechanical clock with a pendulum mechanism in 1657. In 1921, W. H. Shortt invented a pendulum clock so accurate that it was used in the Royal Observatory in Greenwich. The measurement standard for time is the atomic clock. These clocks measure time so accurately, they will not gain or lose more than a second in 300 years.

TICK TOCK

2. Attach the clock face, and push a piece of dowelling through at the bottom, as shown.

1. Cut the top off a large, square box, as shown. This will be the base of the clock. Cut a second piece of cardboard, the same width as the box, to be the clock face.

3. Decorate your clock face and design and draw a dial. Mark the indicator scale when you can observe the size of the pendulum swing.

6. Make an arrow to be the hand that points to the dial. Pin the pendulum to the piece of dowelling so it can swing freely.

4. The pendulum can be any length you wish. Use a dowel with a circle of cardboard attached. To give weight to the bob, put modeling clay behind the card.

5. Balance the pendulum by fixing a short length of dowelling, with a circle of cardboard at each end, across the top, as shown.

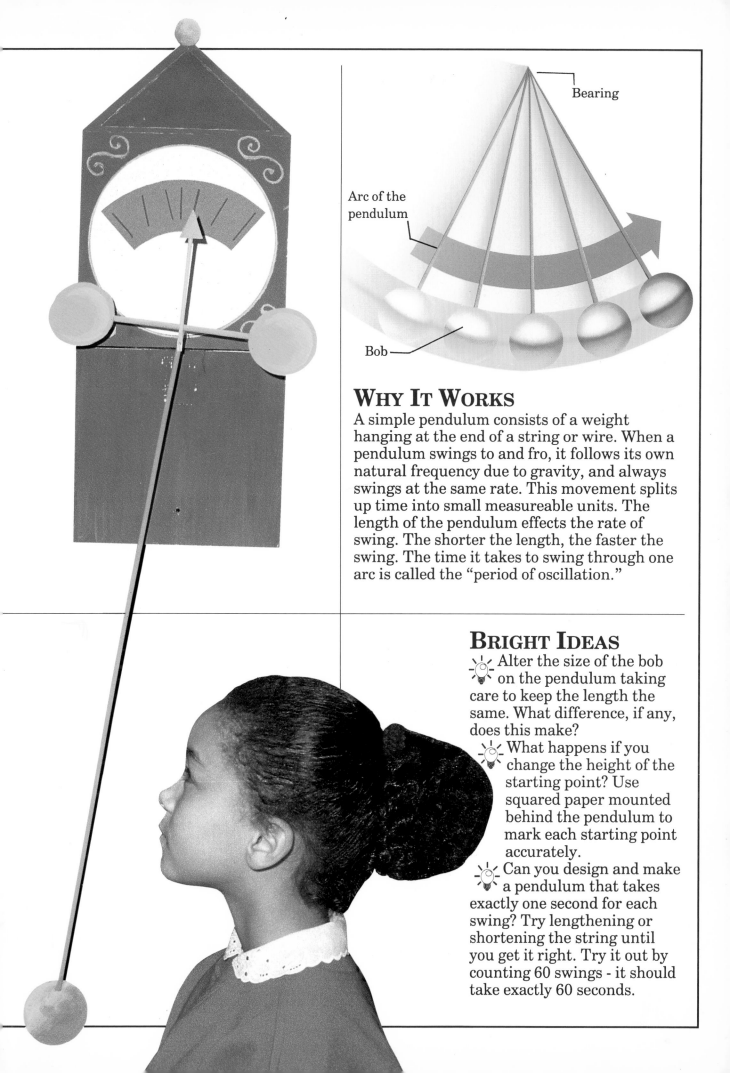

Bearing

Arc of the
pendulum

Bob

WHY IT WORKS

A simple pendulum consists of a weight
hanging at the end of a string or wire. When a
pendulum swings to and fro, it follows its own
natural frequency due to gravity, and always
swings at the same rate. This movement splits
up time into small measureable units. The
length of the pendulum effects the rate of
swing. The shorter the length, the faster the
swing. The time it takes to swing through one
arc is called the "period of oscillation."

BRIGHT IDEAS

Alter the size of the bob
on the pendulum taking
care to keep the length the
same. What difference, if any,
does this make?

What happens if you
change the height of the
starting point? Use
squared paper mounted
behind the pendulum to
mark each starting point
accurately.

Can you design and make
a pendulum that takes
exactly one second for each
swing? Try lengthening or
shortening the string until
you get it right. Try it out by
counting 60 swings - it should
take exactly 60 seconds.

REGULATING TIME

In 1808, Sir William Congreve designed the rolling ball clock. The timing mechanism was a ball that took exactly 15 seconds to roll, in a zigzag fashion, in grooves along an inclined ramp. An accurate pocket watch was only possible after Robert Hooke discovered, in 1685, that a vibrating spring could provide a regular rhythm for a watch. The modern watch may use the natural vibrations of quartz crystal, 100,000 times per second, to keep time.

7

7. Add a cardboard collecting channel at the bottom. Every time a marble reaches the bottom, it lands on the seesaw, pushing the straw upward. The straw moves up the plastic tube, knocking the next ball down the slope.

HELTER SKELTER

1

1. Take a rectangle of cardboard, 16in x 14in, and divide into four columns. Cut 6 parallel slots in each column, sloping slightly from left to right.

2

2. Cut a small door, as shown, and score and fold to make this 3-D shape. Tape the joints.

3. Now measure, cut and score 24 cardboard slots to fit into each of the slits. Slot them into position to make a zigzag channel for the marbles to roll along.

4

4. Make a channel for the marbles from cardboard. Cut a marble-sized hole at one end.

5

5. Tape a narrow plastic tube inside. Glue in place, raising it up at one end.

6. The seesaw mechanism at the bottom is a piece of cardboard, weighted underneath. A piece of tubing acts as the pivot and a straw runs from one end up into the plastic tube above.

6

WHY IT WORKS

The rolling marble (1) is pulled down the slope by gravity. The force of gravity causes the marbles to accelerate as they fall. The marbles slow down when they are forced to change direction. Each marble takes exactly the same amount of time to roll down the slope. If this time is known, it can be multiplied by the number of marbles at the bottom (2) to calculate how much time has passed.

BRIGHT IDEAS

What happens if you use marbles of varying size? Is it possible to time each marble's descent with total accuracy? Does each take exactly the same amount of time? What is the total time taken for all of the marbles to reach the bottom?

Design and make a water clock like the one shown here. Will the flow be constant? How does the water clock compare with the sandglass? Can you design a water clock that takes exactly 10 minutes to empty?

Scale

SPEED AND ACCELERATION

Speed is how fast something is moving. Velocity describes both speed and direction. When a car turns a corner, speed may stay the same, but velocity changes. Speed describes how far an object travels in a period of time. For example, a snail moves at about 0.03 miles per hour, while Concorde travels about 1,300 miles per hour. A speedometer, like the one pictured here, indicates how fast a car is moving. Acceleration is how much the speed increases in a period of time. A decrease in speed is called deceleration.

AT FULL SPEED

1

1. Cut the road from stiff cardboard as wide as a shoe box, but twice as long. Secure a small peice of cardboard in the middle of one side with a paper fastener.

2

2. Tape one end of the road to the narrow end of a shoe box. It can be lifted to different heights.

3

3. Cut out one quarter of a circle. Divide it into angles of 10 degrees, and cut slits along the edge. Attach to the side of the box.

4. Check that the piece of cardboard on the road is in the correct position to slide into a notch. Pierce a hole in the lid of the plastic bottle. Cut off the bottom section. Tape over the hole, invert and fill with paint. Mount on top of the car.

4

WHY IT WORKS

By observing the distance between the drops of paint on the inclined ramps you can estimate the speed of the vehicle. When they are close together, the speed is slowest. If the spaces are uniform, the vehicle must be traveling at a constant speed. If the spaces widen, the vehicle has accelerated, if they narrow it has decelerated. The bigger the mass of an object, the greater the force needed to make it move. When the slope is steeper, the truck accelerates faster. The spaces between the drops are wider apart towards the bottom of the slope. The speed can be calculated by dividing the distance by the time taken.

Force of gravity

Gentle slope

Force of gravity

Steep slope

Wide spaces

BRIGHT IDEAS

Use a variety of toy cars on a sloping ramp and experiment with differing angles. Which car travels furthest? Did you use the same "push" each time to ensure a controlled experiment?

Run the same car down a variety of angles, and allow it to run into a shoe box distance that the box has moved. The distance that it moves depends on the speed.

Try running on a beach at various speeds. Use a stop-watch to time yourself and measure the spaces between the footprints. Your footprints will be further apart, the faster you run.

5

5. Set the angle of the road. Place a long piece of paper over the road to record your results each time. Position the car at the top of the slope. Remove the tape just before it is released. Time each run accurately. Be careful not to push the car.

WEIGHT AND GRAVITY

Weight is due to the forces of gravity. Gravity is the force of attraction that acts toward the Earth's core, pulling us all downward. Sir Isaac Newton (1642-1727) developed the theory of gravitational pull. The Earth is held in orbit around the Sun by the pull of gravity. Gravitational pull is visible daily in the movement of the tides. They are a direct consequence of the gravitational pull between Earth, Moon and Sun. The weight of an object can vary slightly at different places on Earth.

PULL HARD!

1

1. Measure out a vertical indicating scale on a long sheet of cardboard. Space the divisions evenly.

2

2. Cut the base from a plastic bottle. Paint it a bright color. This will be a weighing pan. Pierce 2 holes directly opposite each other in the sides and attach a piece of thread, as shown.

3. Use the painted top of a plastic bottle as a holder from which to suspend the balance. Cut a flap, as shown here, and make two holes in it. These will be the screw holes through which it can be attached to a length of wood.

3

4. Tie an elastic band to the thread already attached to the weighing pan. Loop the free end to the plastic holder. When suspended the weighing pan must be level. If it is not, adjust the elastic.

4

5

5. Position the wood in a secure, vertical position. Attach a pointer to the base of the weighing pan. This will simplify the reading of the scale. Establish the position of the pointer when the pan is empty.

6

6. Mark the point on the indicating scale where the base of the empty pan rests. All other measurements must be taken from that point. Select a variety of objects to weigh. Place each object inside the weighing pan and read from the scale when the pan is still.

WHY IT WORKS

Robert Hooke (1635-1703) devised the theory of elasticity. Hooke's Law states that the extension of a spring is directly proportional to the force applied to it. You observed how much the elastic stretched each time you weighed an object on the "spring" balance. When the weight doubles or trebles, the elastic stretches proportionally. This is because the gravitational pull is stronger for objects containing a greater amount of matter and therefore their weight is greater.

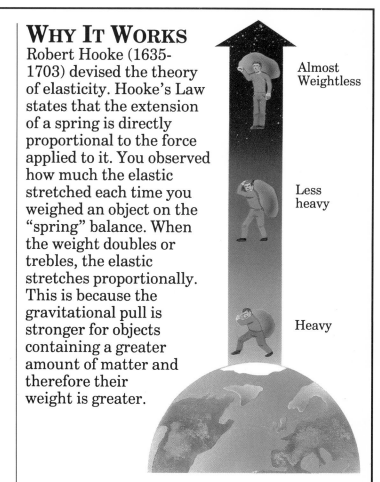

Almost Weightless

Less heavy

Heavy

BRIGHT IDEAS

🔆 Collect a variety of objects to weigh on a real spring balance. Record the length of the spring in inches. Chart these results and look for a relationship between them. The greater the force recorded, the further the spring is likely to stretch. Does the size of an object make a difference to the force of gravity acting on it?

🔆 Suspend a stone from a piece of elastic. Now suspend the same object from the elastic and lower it into a bucket of water until it is immersed. What do you observe? Why does this happen?

🔆 Tie a long length of string around the middle of a heavy, closed book. Now lift the book by holding the string at each end. Try to make the string horizontal by pulling hard. It is impossible! The force pulling down on the book and the angle created between the two pieces of string makes it very hard to lift.

MEASURING MASS

Mass is not the same as weight. Mass is the amount of matter in an object, measured in kilograms – it is constant wherever the object is. Each civilization developed its own system of weight measurement based on mass. A Roman pound weight had the same mass as one eightieth of a cubic foot of water. Nowadays, weighing has been standardized by the introduction of metric units of mass – kilograms and grams. A bar of gold has a greater mass than a piece of wood of the same size. Comparing the mass of different materials can be done using a balancing scale.

BALANCING ACT

1

1. Take two plastic beakers. In the top of each beaker make two holes directly opposite each other. Thread some string through the holes, leaving the same length free on each side.

3

2

2. On a sheet of cardboard measure a horizontal indicating scale. Be as accurate as possible. The distance between each mark on the scale must be the same. Make the markings clear so that they can be read easily.

3. Choose a strong coat hanger that has a rigid structure. Decorate the hanger. Cut one end of a plastic straw to make a point. Suspend the straw from the center of the hanger. Ensure that it hangs vertically.

4

4. Suspend the hanger from a hook. Position the indicating scale so that the straw is in line with the center mark. Hang both beakers, one from each end of the hanger. Check that they are level.

3

5

WHY IT WORKS
Balancing scales work on a system of levers. When the weight on both sides is the same, the indicator will point to the center of the scale. If the indicator points to the right side of the scale, the object in the left-hand container has the greater mass.

Scale

5. You can use marbles as weights. Ensure that they are of identical mass. Each time you weigh an item, record the number of marbles required to maintain the balance of the scales. If the straw points to the right of the center mark, more marbles must be added. When it points to the left, marbles must be removed.

BRIGHT IDEAS
Find the mass of a tennis ball and a baseball, or a cricket ball of the same size. Collect objects of similar size (volume). Estimate their mass, then weigh them. What is the difference between mass and volume?

Tennis ball

Cricket ball

AREA, VOLUME, AND CAPACITY

The first crude tools of measurement of length were parts of the body. Shorter units of length were measured against the thumb – the Roman "uncia" became the inch – the handspan or the foot. Longer lengths were measured against the forearm or a stride. From direct measurements of length, height and depth, indirect measurements of area, volume and capacity can be calculated. The modern unit of measurement for a large area of land is the hectare. A hectare equals 100 acres or 10,000 sq. meters. In 1794, the Cadil was established as the metric standard of capacity – it became known as the liter.

SURFACES AND SPACES

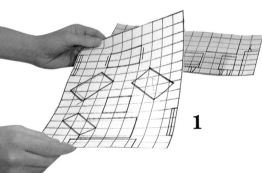

1. Use a yardstick or a tape measure to take accurate measurements of a room and the furniture. You need to know all dimensions. Scale down the measurements and transfer them to a piece of graph paper.

2. Each piece of furniture is drawn to scale on graph paper. Use these as patterns to draw round on cardboard. Score along fold lines to achieve a neat 3-dimensional shape.

3. Construct a scale model of three walls and the floor of the empty room. Draw a net, with flaps, to scale, on graph paper, then transfer it onto cardboard. Cut out the windows and glue together.

4. Arrange the furniture inside the room, so that it looks like the original room from which you took measurements.

1in
Length 1in
Volume 1in³

1in
Length 2in
Volume 8in³

1in
Length 3in
Volume 27in³

1in
Length 4in
Volume 64in³

1in
Length 5in
Volume 125in³

WHY IT WORKS

Area is the size of a surface; it is commonly measured in square units. The area of a rectangular shape is calculated by measuring length times width.

Volume is the amount of space taken up by a shape. Capacity is the amount, or volume, that a container can hold. All are indirect measurements of size and space. A three-dimensional shape has length, height and depth. The volume of a regular 3-D shape, like a cube, can be measured by finding the product of these three measurements.

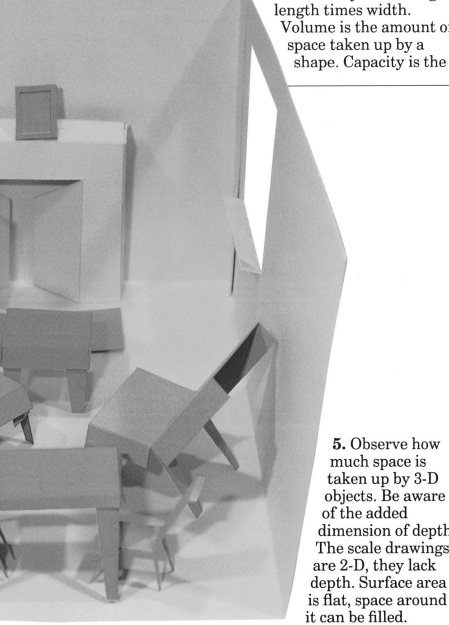

5. Observe how much space is taken up by 3-D objects. Be aware of the added dimension of depth. The scale drawings are 2-D, they lack depth. Surface area is flat, space around it can be filled.

BRIGHT IDEAS

Make a tangram puzzle (below) by cutting up a square. See how many shapes you can make using all 7 pieces each time. Will the area be the same each time?
Use 1in plastic cubes to build a series of larger cubes. How many cubes are used on each face? What is the area of one face? What is the total surface area? What is the volume of each cube? What is the relationship between volume and the surface area of one face?

DENSITY

Britain covers an area of about 95,000 sq. miles. France covers an area of approximately 210,000 sq. miles – an area twice as large. A similar number of people live in each. Therefore, France has a lower population density because it has twice as much space for the same number of people. Similarly, the density of a solid or a liquid can only be measured indirectly. It is dependent upon knowing the mass and the volume. Salt water is more dense than fresh water. This is why it is easier to float in the sea, particularly the Dead Sea.

HOW DENSE?

1. Cut a length of straw and, to one end, attach a ball of modeling clay. Place it gently in water to ensure that it is balanced.

2. First, measure the density of water. Measure out a given amount of water and pour it into a transparent beaker. Place the hydrometer in the water and mark the straw at the surface.

BRIGHT IDEAS

💡 Collect a variety of objects made from different materials. For example, you could use wood, cork, stone, glass, plastic, and metal. Place them in a large container of water. Do they float or sink? An object with a lower density than water will float. Those with a higher density will sink.

💡 Calculate the density of water. Find the mass of an empty beaker. Fill the beaker with a specified volume of water. Find the combined mass of the beaker and the water, then subtract one from the other to find the mass of the water. Use the formula density equals mass divided by volume.

3. Measure out the same quantity of oil and pour it into an identical beaker. Place the hydrometer in the oil and observe the position of the mark on the straw. Repeat for other liquids e.g. vinegar, salt water, and sugar water.

3

WHY IT WORKS

You have made an hydrometer to measure the relative density or specific gravity of various liquids. This means that each measurement is compared to the measurement taken for water. The more dense a liquid is, the easier it is for objects to float. In a liquid less dense than water, the straw sinks lower. The lightness or heaviness of a given volume of any substance is called its density. Density is dependent upon how tightly the molecules of each substance are packed. If various liquids of different densities are gently poured into a transparent container beginning with the most dense first, they will settle in colored bands according to the density of each. Liquids you can use include salad oil, salt water, and colored fresh water.

— Least dense

Most dense

SURVEYING THE LAND

Tax gatherers in Ancient Egypt were probably the first professional surveyors. They developed a system called triangulation because it was easier to measure angles accurately than it was to measure distances. Surveyors still use the system of triangulation today – they can even use it to measure the height of mountains. The measuring tool used is the theodolite, which is mounted on a tripod like the one pictured here. Thales (624-546 BC), a Greek merchant and teacher, devised a method for measuring the height of the Great Pyramid in Ancient Egypt.

WHAT ANGLE?

1

1. On a piece of cardboard, draw a circle, with a radius of 12in and, with the help of an adult, cut it out with a matte knife. Use a cutting board. Rule a diameter across the circle, then measure and mark divisions of 10 degrees all the way round. There will be 36 divisions, a circle has 360 degrees. Cut the circle in half.

2. Mark the divisions clearly. Make two "viewers" from rolls of cardboard. Glue them between the two semi-circles, at either end.

2

3. Wind a strip of cardboard round a heavy object, such as a battery. Attach this to one end of a length of string at least 24 in long. Secure the other end to the center of the straight edge of the theodolite.

3

WHY IT WORKS

When you viewed a landmark along your theodolite, the plumb line indicated the size of the angle at each end of the base line of an imaginary triangle. Once the distance and the vertical angle is known between two points, a surveyor can calculate the difference in height. That measurement is added to the height of the theodolite to give the total height of the landmark.

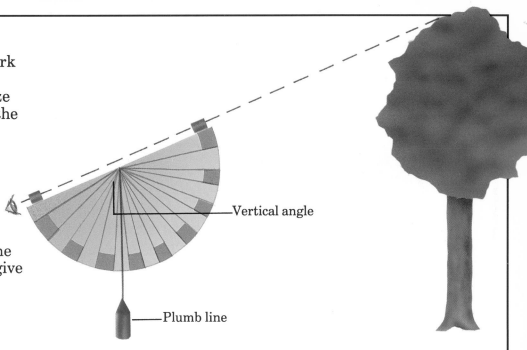

Vertical angle

Plumb line

4. To measure an angle, sight a particular point on a tall landmark through the viewers. When the plumb line is motionless, ask someone to estimate the size of the angle from the alignment of the string.

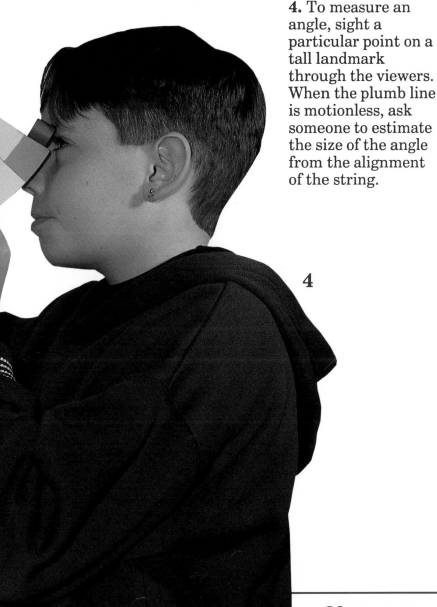

4

BRIGHT IDEAS

You can make another kind of theodolite out of a sheet of cardboard cut to the shape of a right-angled triangle. You will also need a plumb line, a ruler and a tape measure. By lining up the longest side of the cardboard with the top of a tree or pole, so that the plumb line hangs exactly along the short side nearest to it, you can calculate the height of the tree. You will need to walk toward or away from the tree until this happens. Now measure the distance between yourself and the foot of the tree. Its height is that distance plus your own height.

Measure the height of a friend. Ask him to stand beside a tall tree, then move away until he appears to be 2in tall on a ruler. Now measure the height of the tree from the same distance. Can you estimate the height of the tree by working out how many times bigger than your friend the tree is?

READING WEATHER SIGNALS

Earth has weather because it is surrounded by a shallow blanket of air called the atmosphere. The Sun's rays pass through the atmosphere to reach the ground. The tilt of the Earth on its axis, and the curvature of its surface, create the variety of weather patterns and climatic zones experienced all over the world. For hundreds of years people have tried to forecast the weather. This is done by measuring such things as wind speed using an anemometer like this one (right), and monitoring any changes. Atmospheric pressure is measured using a barometer. The mercury barometer was invented in 1643 by Torricelli, an assistant of Galileo. The first weather station was built for Napoleon III in 1854.

STORMY WEATHER

1. Take an empty box and remove the lid. Decorate it appropriately with weather symbols.

1

2

2. Measure and draw a curving indicating scale as shown. Make sure the divisions are equal. Use a marker pen to ensure clarity. Indicate "high" at the top of the scale, and "low" at the bottom.

3. Cover the box with plastic film, particularly the open top. Ensure that the plastic is taut and crease-free across this open part. This is the "drum" of your barometer.

3

4

4. Attach the indicating scale to the back of the box. Make sure that it protrudes beyond the box.

5

5. Secure the plastic straw horizontally to the scale card with a pin. Direct one end of the straw to the center of the scale. Fix the other pin inside as a pointer.

WHY IT WORKS

This barometer works as the air pushes down on the cellophane. When air pressure changes, the surface of the cellophane moves up or down. If air pressure falls the pointer falls – rainy weather is approaching. A rise in air pressure means the pointer will rise, indicating dry weather. The anemometer reacts to wind speed. If the wind is strong, it will push the strip of cardboard further up the scale.

Indicator

Air pressure

Scale

Scale

Wind

BRIGHT IDEAS

1. Design and make a portable anemometer. Use a protractor and compass to draw a scale on a piece of thick cardboard.

1

2

2. Tape half a table tennis ball to a strip of cardboard and cut a window so you can read the scale.

3

6. Use sticky tape to attach the thread to the straw. Attach the free end of the thread to the surface of the plastic. The straw must be horizontal.

3. Pin the strip to the card. The scale should be:

Angle:	80	60	40	20
mph	8	15	21	32

Scientific Terms

ANEROID BAROMETER An instrument used to measure air pressure.

ASTRONOMY The scientific study of the Universe.

CONSTELLATIONS Groups of stars visible in the night sky, recognized easily from their familiar patterns.

FRICTION The force that resists movement when one surface moves relative to another.

GRAVITY The force of attraction between two objects. The more massive an object is, the greater its force of gravity.

GREENWICH MEAN TIME Standard time throughout most of the world, calculated from the Prime Meridian at Greenwich, London.

BRITISH IMPERIAL MEASUREMENTS Standard measurements used in Great Britain before the introduction of the Metric System.

LUNAR CYCLE The time taken for the Moon to make one complete revolution of the Earth - approximately 29 days 12 hours 44 minutes 3 seconds.

METEOROLOGIST A person who undertakes the study of weather forecasting.

METRIC SYSTEM The decimal system of units based on the meter.

POLE STAR The star closest to the north celestial pole. It appears to remain fixed in the sky.

SURVEYOR A person who plots detailed maps of the land.

TANGRAM A Chinese puzzle comprising seven cut pieces.

URSA MAJOR A constellation in the Northern hemisphere.

Index